MACAQUE MONKEY AS PET

What You Need to Know

A COMPLETE GUIDE TO THEIR HABITAT, CARE, DIET, OWNERSHIP, BEHAVIOR, INTELLIGENCE, BREEDING AND MANY MORE

DR MORRIS HART

Table of Contents

Introduction to Macaque Monkeys

Macaque monkeys, which belong to the genus Macaca, are a diverse group of primates found around the world. They are noted for their intellect, flexibility, and complex social activities, which make them both appealing and challenging as potential pets. In this detailed tutorial, we will look at macaque monkeys' natural habitat, physical qualities, social dynamics, and the factors to consider when keeping them as pets.

Natural Habitat and Distribution

Macaque monkeys are native to Asia, North Africa, and Gibraltar, where they live in a variety of habitats ranging from tropical rainforests to hilly regions and cities. Their versatility has enabled certain species to thrive even in human-altered habitats, where they frequently interact with human populations.

Geographic Distribution

The genus Macaca contains approximately 23 known species, each adapted to a unique ecological niche within their particular habitats. The Rhesus macaque (Macaca mulatta) can be found throughout South, Central, and Southeast Asia, but the Japanese macaque (Macaca fuscata), often known as the snow monkey, lives in Japan's colder regions and has adapted to life in snowy locations.

Physical characteristics

Macaques vary greatly in size and appearance according to the species. They have powerful bodies with long limbs, a short tail (or no tail at all in certain species), and a strong sense of balance, which allows them to easily cross complicated settings like trees and cliffs.

Size and Weight

Macaques range in size from M, which is relatively little. fascicularis, which can weigh between 5 and 8 kilograms (11 to 18 pounds), to the bigger M. Sylvanus can weigh 15 kilos (33 pounds) or more.

Fur Colors and Textures

Their fur can range in color from gray or brown to reddish or even golden, with unique face markings or patterns that differ between species. Their fur texture can vary from coarse to soft, depending on their habitat and environment.

Social Structure and Behavior

Macaque monkeys are highly sociable animals that form intricate hierarchical groups known as troops. These troops are often made up of numerous females and their progeny, as well as a dominant male or multiple males vying for leadership and breeding opportunities.

Troop Dynamics

Within a troop, social interactions are critical for sustaining order and collaboration. Dominance hierarchies are formed by a variety of social behaviors, including aggression, grooming routines, and vocalizations. Subordinates frequently exhibit submissive actions to avoid disagreement with higher-ranking members.

Communication

Macaques communicate using a variety of vocalizations, including calls, screams, and grunts, as well as nonverbal signs such as facial expressions, gestures, and body postures. These modes of communication aid in sustaining social bonds, warning danger, and coordinating group activities such as hunting and grooming.

Diet and Eating Behavior

Macaques are omnivores, which means they eat a wide variety of foods based on their environment and availability. Their diet mainly consists of fruits, seeds, leaves, insects, tiny vertebrates, and, on rare occasions, eggs or small animals.

Foraging Strategies

Foraging behavior varies by species and is governed by factors including food availability and group social dynamics. Some macaques are particularly opportunistic, raiding farms or scavenging from human settlements, but others rely on natural resources in their native habitats.

Dietary adaptations

Certain animals have evolved unique adaptations for obtaining food, such as cheek pouches for storing seeds or a specific manner of cracking nuts with pebbles or tools. These adaptations showcase their cognitive capacity and aptitude to create in the face of environmental difficulties.

Cognitive Ability and Tool Use

Macaque monkeys are well-known for their cognitive flexibility and ability to use tools, which was previously thought to be unique to humans and a few other primate species. This ability varies by species, but it is especially well documented in societies where people utilize rocks or sticks to get food or manipulate items.

Research and insights

Studies on tool usage in macaques have shed light on monkey cognition, social learning, and the evolutionary origins of human technological ability. Observations of wild and captive populations continue to reveal new features of their problem-solving abilities and the cultural transmission of information within communities.

Conservation Status & Threats

Many macaque species suffer serious challenges to their survival, including habitat degradation, fragmentation, human-animal conflict, and illegal wildlife trading. Conservation initiatives are crucial for protecting people and maintaining natural environments for future generations.

Threats To Habitat

Deforestation, urbanization, and agricultural development are all direct threats to macaque populations, decreasing their access to food and upsetting their social structures. Conservation organizations strive to create protected areas and corridors that connect fragmented habitats, allowing populations to migrate and thrive.

Wildlife Trade

Despite international rules and attempts to prevent wildlife trafficking, the illegal capture and exchange of macaques for pets, study, and traditional medicine continues to have an influence on wild populations. Education and enforcement of regulations are critical for reducing demand and safeguarding vulnerable species from exploitation.

Ethical Considerations for Pet Ownership

The decision to keep a macaque monkey as a pet raises ethical concerns that go beyond the practical issues of care. Primates, notably macaques, have intricate social and psychological requirements that are difficult to meet in a home setting.

Welfare Concerns

Primates are highly sociable animals that require social connection, mental stimulation, and opportunity to engage in natural behaviors like climbing, foraging, and grooming. Pet macaques' confinement and seclusion can cause stress, behavioral disorders, and physical health concerns if their requirements are not handled adequately.

Legal Restrictions

Many nations have strict regulations or outright bans on owning primates as pets, including macaques, due to welfare concerns and potential threats to human health and safety. legislation and regulations vary greatly, therefore potential owners should extensively research local legislation before adopting a macaque as a pet.

Macaque monkeys are extraordinary organisms with a long evolutionary history and sophisticated social behaviors. While their intellect and adaptability make them attractive candidates for research, keeping a macaque as a pet necessitates careful consideration of their unique needs, legal constraints, and ethical consequences. Understanding their natural habits, food needs, and social dynamics allows us to better appreciate and protect these amazing primates in both wild and captivity environments.

Chapter 1

Legal and Ethical Considerations for Owning Macaque Monkeys as Pets

Keeping a macaque monkey as a pet requires navigating complex legal structures and ethical concerns. Primates, particularly macaques, are highly intelligent and gregarious creatures with unique care requirements that are frequently difficult to provide in a household context. This detailed handbook examines the legal limits, ethical difficulties, welfare implications, and sociological consequences of keeping a macaque monkey as a pet.

Legal Framework and Regulations

Many countries impose strict laws on the ownership and trading of macaque monkeys due to concerns about animal welfare, public safety, and conservation. Laws concerning monkey ownership differ greatly between nations, impacting whether and how people can lawfully keep macaques as pets.

International Regulations
At the international level, the Convention on International Trade in Endangered Species of Wild Fauna and Flora (CITES) governs the trade and conservation of endangered species, including certain macaques. Certain macaque species are listed under Appendix II of CITES, and their import and export require licenses to guarantee that trade does not harm their survival in the wild.

National Legislation

National regulations also govern the legality of keeping macaques as pets. For example, in the United States, primat ownership is governed by the federal Animal Welfare Act (AWA), which establishes guidelines for their care and shelter at research centers and for display reasons. However, state regulations on private possession differ, with some explicitly prohibiting primates as pets due to concerns about public safety and animal welfare.

Local Regulations

Local rules and municipal bylaws may also limit or prohibit the possession of primates in certain jurisdictions. These rules frequently reflect community concerns about the possible risks connected with exotic animal ownership, such as zoonotic infections, violence, and the ability to provide proper care.

Ethical Considerations for Pet Ownership

The decision to keep a macaque monkey as a pet raises serious ethical concerns about animal welfare, conservation, and the fundamental rights of nonhuman animals. Primates have advanced cognitive ability, social activities, and emotional capacities that are difficult to imitate in captivity.

Social and Psychological Needs

Macaque monkeys are extremely social primates with complex social structures and a variety of communication, grooming, and play behaviors. In the wild, they spend a lot of time foraging, investigating their surroundings, and engaging with conspecifics. Captivity can severely limit their natural behaviors, resulting in tension, boredom, and behavioral disorders if their social and environmental demands are not addressed.

Cognitive stimulation
Primates, particularly macaques, have high cognitive capacities like problem solving, tool usage, and social learning. These intellectual capacities necessitate mental stimulation and enrichment opportunities, which are frequently unavailable in household situations. Without appropriate stimulation, pet macaques may exhibit stereotypical behaviors or hostility, indicating psychological distress.

Longevity and Commitment
Macaque monkeys can live in captivity for several decades, necessitating a long-term commitment to their health and well-being. Potential owners must determine whether they can give lifelong care, such as specialized foods, veterinary care, and enrichment activities that improve physical and mental wellness.

Conservation Implications
The demand for exotic pets, such as macaque monkeys, adds to pressures on wild populations through habitat degradation, illegal trade, and unsustainable harvesting. Conservation organizations highlight the significance of preserving natural habitats and reducing demand for wild-caught animals through education, advocacy, and enforcement of wildlife protection legislation.

Habitat loss and fragmentation
Wild macaques face major risks from habitat loss and fragmentation caused by deforestation, agricultural expansion,

and urban development. Conservation efforts are centered on protecting vital habitats and creating protected areas to support viable populations of these primates.

Wildlife Trafficking

Despite international rules aimed at combating wildlife trafficking, the illegal capture and exchange of macaques for pet purposes continues to have an influence on natural populations. The illegal wildlife trade endangers animal welfare and biodiversity, causing population decreases and jeopardizing the survival of endangered species.

Public Health and Safety Concerns

The possession of macaque monkeys as pets presents public health and safety concerns, as they have the potential to spread zoonotic diseases and exhibit unpredictable behavior. Primates can carry human-transmissible diseases such as viral, bacterial, and parasitic infections, putting both owners and the general public at risk.

Zoonotic diseases

Macaques are susceptible to a variety of zoonotic diseases, including herpes B virus (Cercopithecine herpesvirus 1), which can be lethal to people if transmitted via bites or scratches. Other potential zoonoses include tuberculosis, hepatitis, and the simian immunodeficiency virus (SIV), emphasizing the significance of strict health procedures and veterinary care for pet macaques.

Public Safety Risks

Primates, notably macaques, are physically strong and agile, which can endanger people, especially in unfamiliar or stressful situations. Aggressive behavior, territorial disputes, and dominance hierarchies within groups can result in injuries or conflicts, necessitating careful handling and management procedures.

Responsible Ownership and Education

Given the complexities and difficulty of keeping macaque monkeys as pets, ethical ownership practices and education are critical for increasing animal welfare and mitigating detrimental effects on both captive and wild populations.

Education and Awareness.
Public education campaigns play an important role in raising knowledge about primates' unique needs and the ethical implications of keeping unusual animals as pets. Outreach campaigns, instructional materials, and collaborations with conservation organizations can help potential owners understand the duties and ethical implications of primate ownership.

Alternative Conservation Efforts
Rather than keeping macaque monkeys as pets, donating to conservation organizations and wildlife sanctuaries provides an alternative method to contribute to species conservation and wellbeing. Sanctuaries provide refuge for rescued primates, rehabilitate them for possible release back into the wild, and educate the public about the need of protecting natural environments.

Finally, the decision to keep a macaque monkey as a pet necessitates careful consideration of legal constraints, ethical considerations, welfare implications, and conservation consequences. Primates, including macaques, are complex beings with unique requirements that are difficult to meet in captivity. Understanding and appreciating their natural behaviors, cognitive capacities, and conservation status allows us to make educated decisions that prioritize animal welfare while also contributing to the protection of these extraordinary species in both wild and confined settings.

Chapter 2

Understanding macaque behavior

Macaque monkeys, who belong to the genus Macaca, exhibit a variety of behaviors that are influenced by their evolutionary history, social structures, and environmental adaptations. Studying macaque behavior reveals important information on primate cognition, social dynamics, communication, and adaptation techniques. This comprehensive guide delves into the numerous aspects of macaque behavior, including as social organization, communication, foraging methods, reproductive habits, and interactions with people.

Social Organization and Group Dynamics

Macaque monkeys are highly sociable animals that form intricate hierarchical groups called troops. Understanding their social organization reveals insights about how people interact, communicate, and achieve authority in their groups.

- Troop Structure

Troops usually consist of several people, including adult men, females, and their progeny. A troop's size varies according to species and environmental variables, ranging from a few individuals to several dozen members in bigger groupings.

- Dominance Hierarchy

Competitive interactions and dominance displays between males and females build social hierarchies within macaque troops. Dominant people have first access to resources including food, mates, and resting areas, whilst inferior members may defer to higher-ranking individuals to prevent conflict.

- Matrilineal bonds

Female macaques frequently create matrilineal family networks within the group, resulting in close relationships with their female relatives. These ties are vital for social support, cooperative behavior, and offspring protection, all of which help related individuals survive and reproduce.

Communication and Vocalization

Macaques communicate using a variety of vocalizations, facial expressions, gestures, and body postures to convey information about their social standing, reproductive condition, threat levels, and affiliative connections.

- Vocal repertoire.

Macaques emit a wide range of vocalizations, including shouts, screams, grunts, and coos, each serving a specific communicative purpose. Alarm sounds, for example, warn group members of impending hazards, such as predators, but affiliative calls strengthen social relationships during grooming or resting periods.

- Facial Expressions and Gestures

Macaque facial expressions reflect emotions such as fear, anger, surrender, and calm. Gestures including grooming invitations, threat displays, and play behaviors help to foster social interactions and keep groups together.

- Nonverbal Communication

Nonverbal indicators such as body posture, eye contact, and physical contact provide layers of communication between macaques. These tiny signals aid in establishing social hierarchies, resolving disagreements, and coordinating group activities such as hunting or grooming.

Foraging Strategies and Dietary Adaptations

Macaque monkeys are opportunistic omnivores with variable dietary choices and foraging tactics that differ by species and location. Their capacity to exploit a diverse range of food sources demonstrates their adaptability and cognitive capacities.

- Dietary Preferences

Macaques eat fruits, seeds, leaves, insects, small vertebrates, and occasionally eggs or small animals. Their nutritional choices may change depending on the season or the availability of food in their natural habitat.

- Tool Use and Innovation

Some macaque species employ tools and are innovative in their hunting behaviors, such as cracking nuts with stones or extracting insects from crevices with sticks. The usage of tools displays their ability to solve problems and adapt to environmental difficulties.

- Food Sharing, Social Dynamics

Food sharing among macaques supports social functions by strengthening individual ties and preserving group cohesion. Dominant individuals may have preferred access to food supplies, whilst subordinates benefit from affiliative behaviors and cooperative foraging techniques.

Reproductive Behavior and Parental Care

Social hierarchies, mating patterns, and parental investment all influence macaque reproductive strategies, which represent adaptations to maximize reproductive success and offspring survival in unpredictable contexts.

- Mating Strategies

Mating behavior varies among macaque species, with dominant males frequently controlling access to reproductive females through aggressive displays or social alliances. Females may have mate selection preferences depending on male dominance rank, genetic fitness, or social compatibility within their colony.

- Parental Investment

Female macaques place a high value on maternal care, providing safety, grooming, and access to troop resources for their children. Male engagement in parental care varies by species, with some males helping to provide for their children or protecting them from potential hazards.

- Infant Development

During early growth, infant macaques rely heavily on their mothers for nutrition, warmth, and social learning. Play behaviors, environmental exploration, and interactions with peers all help to promote cognitive growth and social skills, which are necessary for survival.

Conflict between humans and wildlife

Macaque monkeys routinely interact with humans in urban, rural, and natural environments, posing difficulties and opportunities for cohabitation, conservation, and management measures.

- Urban Adaptation

Some macaque species have adapted to urban surroundings, utilizing anthropogenic food sources and structures while avoiding possible hazards such as vehicle collisions, domestic pet predation, and disputes with human occupants.

- Crop raiding and human-wildlife conflict

Macaques may raid agricultural crops in quest of food, causing confrontations with farmers and financial losses. Mitigation measures such as crop protection, habitat restoration, and community engagement aim to lessen human-wildlife conflict while encouraging long-term coexistence.

- Conservation & Management

Conservation activities are aimed at minimizing human impacts on macaque populations through habitat conservation, wildlife behavior education, and the development of ways to decrease unfavorable interactions between humans and monkeys.

- Cognitive Ability and Learning

Macaques have high cognitive capacities like problem solving, tool use, social learning, and cultural transfer of knowledge in group situations.

- Problem-solving skills

In laboratory and field trials, macaques display the ability to solve complicated issues, adjust to fresh difficulties, and employ flexible learning processes.

Tool Use and Innovation

The usage of tools by macaques illustrates their ability to innovate and pass down cultural behaviors through generations. Observations of tool use in wild populations demonstrate the variety of approaches used to get food or alter things in their surroundings.

Social Learning and Cultural Traditions

Social learning is important in macaque behavior because it allows individuals to acquire knowledge, skills, and behaviors by seeing, imitating, and interacting with others. Cultural traditions can develop within groups, altering foraging tactics, communication patterns, and social standards throughout time.

Conservation and Ethical considerations

Understanding macaque behavior helps to improve conservation methods for safeguarding wild populations, preserving natural habitats, and encouraging prudent management techniques to reduce human impacts on monkey species.

- Conservation Challenges

Macaques suffer risks such as habitat loss, fragmentation, illegal wildlife trade, and human-animal conflict, emphasizing the significance of conservation efforts to protect their populations and ecosystems.

- Ethical considerations

The decision to study or engage with macaque populations is based on ethical issues such as animal care, research protocols, and the conservation consequences of human activities on wild monkey species.

Finally, studying macaque behavior reveals important information about these extraordinary primates' ecological, social, and cognitive adaptations. Macaques display complex behaviors that are shaped by evolutionary forces and environmental dynamics, ranging from social structure and communication to feeding methods, reproductive activities, and human interactions. By studying and protecting macaque populations, we can get a better knowledge of primate behavior, encourage animal conservation, and facilitate long-term cooperation between humans and nonhuman primates in a changing world.

Chapter 3

Housing for Macaque Monkeys

Providing appropriate habitat for macaque monkeys is critical to their well-being in captivity. Macaques, like all primates, have unique needs for space, enrichment, social interaction, and environmental stimulation that closely resemble their natural habitat. This comprehensive reference looks at macaque monkey housing needs, including cage design, environmental enrichment, social dynamics, veterinary issues, and regulatory requirements for responsible ownership.

Enclosure Design and Size

The design and size of macaque monkey enclosures should take into account their physical health, psychological well-being, and natural habits. Individuals have enough space to move freely, exercise, socialize, and engage in species-typical activities.

- Minimum Space Requirements

The minimal area requirements for housing macaque monkeys varies depending on species, age, sex, and group dynamics. As a general rule, cages should have enough space for climbing structures, resting places, feeding stations, and enrichment activities.

- Indoor versus Outdoor Enclosures

Enclosures can be created for indoor-only or indoor-outdoor use, depending on the environment, security considerations, and local requirements. Outdoor enclosures should have shaded places, natural foliage, and environmental enrichment to encourage natural behaviors and exposure to natural sunshine.

- Climbing Structures and Perches

Macaques are arboreal monkeys that spend a lot of time climbing, swinging, and exploring their surroundings. Enclosures should have durable climbing structures, perches, ropes, and platforms of varied heights to promote physical activity and natural behaviors.

- Environmental Enrichment

Captive macaque monkeys require environmental enrichment to stimulate their minds, exercise their bodies, and connect socially. Enrichment activities should be customized to the species' unique habits, cognitive abilities, and preferences.

- Behavioral Enrichment

Behavioral enrichment activities include those that encourage natural behaviors including foraging, exploration, grooming, and social interactions. Puzzle feeders, foraging boxes, and food-scattering devices stimulate macaques to solve problems and manipulate objects in order to get food rewards.

- Cognitive Enrichment

Cognitive enrichment activities test macaques' cognitive capacities with tasks like item manipulation, tool use, and memory games. Positive reinforcement training can improve learning, develop mental agility, and strengthen caregiver-primate relationships.

- Sensory Enrichment

Sensory enrichment entails allowing macaques to experience a variety of sights, sounds, smells, and textures that mimic their natural habitat. Visual barriers, aural stimulation, scent trails, and tactile materials enhance sensory experiences and encourage exploratory behavior.

Social Dynamics and Group Housing

Macaque monkeys are extremely sociable creatures that form hierarchical groups with intricate social ties, dominance hierarchies, and affiliative behaviors. Group housing allows for social interaction, grooming, play, and cooperative behaviors, all of which are necessary for psychological well-being.

- Group Composition:

To reduce violence and enhance peaceful relationships within the troop, group composition should take into account aspects like as age, gender, temperament, and social history. Introducing new people gradually and monitoring social dynamics might help to develop stable group structures and avoid conflicts.

- Solitary Housing

Solitary living may be required in some circumstances for macaques exhibiting aggression, stress-related behaviors, or medical issues that necessitate customized care. Solitary cages should nonetheless provide for socialization via visual, auditory, and olfactory contact with conspecifics.

- Quarantine and Integration

Newly acquired macaques should be quarantined to prevent the spread of infectious diseases and to enable time for veterinary examination. To ensure compatibility and reduce stress, it is necessary to prepare carefully, introduce people gradually, and watch their social interactions.

Temperature and Climate Control

Maintaining optimum temperature and humidity levels is critical for macaque monkeys' health and comfort, particularly in areas with strong seasonal fluctuations or indoor situations lacking natural ventilation.

- Temperature Range

Enclosures should maintain a temperature range that is appropriate for macaque species' natural habitat preferences, which is normally between 18°C and 28°C (64°F to 82°F). Heating and cooling systems, thermal gradients, and shaded spaces all help to manage temperatures and give thermal comfort throughout the year.

- Humidity Levels

Humidity conditions should be similar to the natural environment of macaque species, ranging from 30% to 70% relative humidity. Monitoring humidity levels and providing access to water sources, misting systems, or humidifiers can help confined macaques stay hydrated and avoid respiratory problems.

- Lighting and Photoperiod

Natural light cycles regulate circadian rhythms, reproductive behaviors, and physiological functions in macaque monkeys. Natural daylight exposure should be used in enclosures, augmented with artificial illumination to give constant photoperiods and UVB radiation, as needed.

- Natural Daylight

Outdoor enclosures should provide macaques with natural lighting cycles, which influence activities such as feeding, resting, and social interactions. Natural sunlight stimulates vitamin D synthesis, improves visual acuity, and promotes overall health.

- Artificial lighting

Indoor enclosures may necessitate artificial lighting systems to maintain constant photoperiods, imitate dawn and dusk transitions, and deliver UVB sunlight required for calcium metabolism and bone health in captive macaques. Lighting fixtures should be positioned to reduce glare and provide consistent illumination throughout the enclosure.

Feeding stations and Dietary Management

Access to good foods and adequate feeding stations encourages healthy eating habits, reduces competition for food resources, and helps macaque monkeys meet their nutritional needs in captivity.

- Nutritional requirements

Macaque diets should reflect their natural dietary preferences, with a well-balanced mix of fruits, vegetables, leafy greens, protein sources, and vitamins as indicated by veterinary nutritionists. Portion control and feeding regimens help to prevent obesity, nutritional deficits, and digestive issues.

- Feeding stations

Multiple feeding stations located throughout the cage eliminate competition and let all group members to obtain food at the same time. Elevated platforms, hanging feeders, and foraging gadgets enable macaques to feed while also increasing physical activity.

Veterinary Care and Health Monitoring

Regular veterinary checkups, preventive healthcare measures, and health monitoring techniques are critical for quickly diagnosing and resolving medical disorders in confined macaques.

- Veterinary examinations

Physical assessments, dental evaluations, immunizations, and parasite screens are all part of scheduled veterinary examinations, which help to maintain optimal health and detect early symptoms of illness or damage. Primate veterinarians provide specialist care that is geared to the particular physiological and behavioral demands of macaques.

- Health Monitoring

Macaque health is monitored on a regular basis by assessing body condition, behavioral changes, reproductive status, and fecal samples for symptoms of gastrointestinal health, parasites, or infectious disorders. Individual medical records and health charts capture the health histories and treatment strategies of each macaque under supervision.

Waste Management and Environmental Hygiene

Maintaining clean, sanitary settings reduces pathogen exposure, reduces stress-related behaviors, and promotes general health and well-being in caged macaques.

- Waste Removal

To reduce odor accumulation and prevent bacterial contamination, enclosure designs should include effective waste collection systems like drainable flooring, litter boxes, or substrate modifications. Regular cleaning schedules and sanitation practices provide sanitary surroundings for macaque monkeys and caregivers.

- Environmental Enrichment

Environmental enrichment activities include foraging, exploration, grooming, and social interactions. Puzzle feeders, foraging boxes, and food-scattering devices stimulate macaques to solve problems and manipulate objects in order to get food rewards.

- Cognitive Enrichment

Cognitive enrichment activities test macaques' cognitive capacities with tasks like item manipulation, tool use, and memory games. Positive reinforcement training can improve learning, develop mental agility, and strengthen caregiver-primate relationships.

- Sensory Enrichment

Sensory enrichment entails allowing macaques to experience a variety of sights, sounds, smells, and textures that mimic their natural habitat. Visual barriers, aural stimulation, scent trails, and tactile materials enhance sensory experiences and encourage exploratory behavior.

Social Dynamics and Group Housing

Macaque monkeys are extremely sociable creatures that form hierarchical groups with intricate social ties, dominance hierarchies, and affiliative behaviors. Group housing allows for social interaction, grooming, play, and cooperative behaviors, all of which are necessary for psychological well-being.

- Group Composition:

To reduce violence and enhance peaceful relationships within the troop, group composition should take into account aspects like as age, gender, temperament, and social history. Introducing new people gradually and monitoring social dynamics might help to develop stable group structures and avoid conflicts.

- Solitary Housing

Solitary living may be required in some circumstances for macaques exhibiting aggression, stress-related behaviors, or medical issues that necessitate customized care. Solitary cages should nonetheless provide for socialization via visual, auditory, and olfactory contact with conspecifics.

- Quarantine and Integration

Newly acquired macaques should be quarantined to prevent the spread of infectious diseases and to enable time for veterinary examination. To ensure compatibility and reduce stress, it is necessary to prepare carefully, introduce people gradually, and watch their social interactions.

- Temperature and Climate Control

Maintaining optimum temperature and humidity levels is critical for macaque monkeys' health and comfort, particularly in areas with strong seasonal fluctuations or indoor situations lacking natural ventilation.

- Temperature Range

Enclosures should maintain a temperature range that is appropriate for macaque species' natural habitat preferences, which is normally between 18°C and 28°C (64°F to 82°F). Heating and cooling systems, thermal gradients, and shaded spaces all help to manage temperatures and give thermal comfort throughout the year.

- Humidity Levels

Humidity conditions should be similar to the natural environment of macaque species, ranging from 30% to 70% relative humidity. Monitoring humidity levels and providing access to water sources, misting systems, or humidifiers can help confined macaques stay hydrated and avoid respiratory problems.

- Lighting and Photoperiod

Natural light cycles regulate circadian rhythms, reproductive behaviors, and physiological functions in macaque monkeys. Natural daylight exposure should be used in enclosures, augmented with artificial illumination to give constant photoperiods and UVB radiation, as needed.

- Natural Daylight

Outdoor enclosures should provide macaques with natural lighting cycles, which influence activities such as feeding, resting, and social interactions. Natural sunlight stimulates vitamin D synthesis, improves visual acuity, and promotes overall health.

- Artificial lighting

Indoor enclosures may necessitate artificial lighting systems to maintain constant photoperiods, imitate dawn and dusk transitions, and deliver UVB sunlight required for calcium metabolism and bone health in captive macaques. Lighting fixtures should be positioned to reduce glare and provide consistent illumination throughout the enclosure.

Feeding stations and Dietary Management

Access to good foods and adequate feeding stations encourages healthy eating habits, reduces competition for food resources, and helps macaque monkeys meet their nutritional needs in captivity.

- Nutritional requirements

Macaque diets should reflect their natural dietary preferences, with a well-balanced mix of fruits, vegetables, leafy greens, protein sources, and vitamins as indicated by veterinary nutritionists. Portion control and feeding regimens help to prevent obesity, nutritional deficits, and digestive issues.

- Feeding stations

Multiple feeding stations located throughout the cage eliminate competition and let all group members to obtain food at the same time. Elevated platforms, hanging feeders, and foraging gadgets enable macaques to feed while also increasing physical activity.

Veterinary Care and Health Monitoring

Regular veterinary checkups, preventive healthcare measures, and health monitoring techniques are critical for quickly diagnosing and resolving medical disorders in confined macaques.

- Veterinary examinations

Physical assessments, dental evaluations, immunizations, and parasite screens are all part of scheduled veterinary examinations, which help to maintain optimal health and detect early symptoms of illness or damage. Primate veterinarians provide specialist care that is geared to the particular physiological and behavioral demands of macaques.

- Health Monitoring

Macaque health is monitored on a regular basis by assessing body condition, behavioral changes, reproductive status, and fecal samples for symptoms of gastrointestinal health, parasites, or infectious disorders. Individual medical records and health charts capture the health histories and treatment strategies of each macaque under supervision.

Waste Management and Environmental Hygiene

Maintaining clean, sanitary settings reduces pathogen exposure, reduces stress-related behaviors, and promotes general health and well-being in caged macaques.

- Waste Removal

To reduce odor accumulation and prevent bacterial contamination, enclosure designs should include effective waste collection systems like drainable flooring, litter boxes, or substrate modifications. Regular cleaning schedules and sanitation practices provide sanitary surroundings for macaque monkeys and caregivers.

- Environmental Enrichment

Environmental enrichment is critical for captive macaque monkeys' cerebral stimulation, physical activity, and social engagement. Enrichment activities for macaques should be varied, exciting, and customized to their species-specific behaviors and preferences.

- Behavioral Enrichment

Behavioral enrichment activities include foraging, exploring, grooming, and socializing. Puzzle feeders, in which macaques must manipulate objects to obtain food, mimic natural foraging activities and improve problem-solving abilities. Hanging food puzzles and scatter feeding strategies can both provide mental stimulation and promote physical activity.

- Cognitive Enrichment

Cognitive enrichment consists of activities that test macaques' cognitive capacities, such as object manipulation tasks, memory games, and learning new behaviors through positive reinforcement training. Training sessions not only provide cerebral stimulation but also enhance the link between caretakers and macaques, hence increasing trust and cooperation.

- Sensory Enrichment

Sensory enrichment tries to activate macaques' senses by exposing them to a wide range of stimuli. This includes offering novel scents, textures, sounds, and visual stimuli that are reminiscent of their natural surroundings. Scent trails, natural sound recordings, tactile items such as branches or ropes, and visual boundaries can all enhance their sensory experiences and encourage adventurous activity.

Social Dynamics and Group Housing

Macaque monkeys are extremely social primates who flourish in group situations with complex social hierarchies, affiliative behaviors, and collaborative interactions. Group living enables macaques to engage in natural social behaviors, form social attachments, and preserve psychological well-being.

- Group Composition:

When housing macaques in groups, it is critical to consider age, gender, temperament, and social history in order to improve group cohesion and reduce aggression. Stable group compositions with compatible members promote pleasant social interactions, grooming exchanges, and cooperative behaviors, all of which contribute to overall troop harmony.

- Solitary Housing Considerations

While group dwelling is preferable for most macaque species, solitary housing may be required in some circumstances. Individuals with aggressive, stress-related behaviors, or medical conditions that necessitate specialized care may benefit from temporary or permanent solitary housing arrangements. Even in solitary enclosures, visual, auditory, and olfactory interaction with conspecifics should be provided to avoid social isolation.

- Quarantine and Integration

Introducing additional macaques into established groups necessitates careful preparation, incremental introductions, and thorough observation of social interactions. Quarantine periods are necessary to check the health of new immigrants and prevent the spread of infectious diseases among the group. Visual introductions, barrier separations, and supervised encounters are examples of integration tactics that help to reduce stress and build social hierarchies.

- Temperature and Climate Control

Maintaining ideal temperature and humidity levels is critical for the health, comfort, and physiological well-being of captive macaque monkeys. Thermal gradients, protected places, and climate control systems should be included in enclosures to mimic natural habitat conditions.

- Temperature Management

To avoid heat stress or hypothermia, enclosures should have temperatures appropriate for macaque species, often between 18°C and 28°C (64°F to 82°F). Heating and cooling systems, insulated structures, and shade arrangements all help to keep temperatures stable and give thermal comfort throughout the year.

- Humidity Regulation

Humidity levels should be consistent with macaques' natural environment, ranging from 30% to 70% relative humidity. Monitoring humidity levels and giving access to water sources, misting systems, or humidifiers help to minimize dehydration, respiratory disorders, and skin problems caused by low humidity conditions.

- Lighting and Photoperiod

Natural light cycles regulate circadian rhythms, reproductive behaviors, and physiological functions in macaque monkeys. Natural daylight exposure should be supplemented with artificial illumination to guarantee constant photoperiods and UVB radiation, as needed.

- Natural Daylight Exposure

Outdoor enclosures should allow macaques to experience natural daylight cycles, which influence behaviors including feeding, resting, and socializing. Exposure to natural sunlight stimulates vitamin D production, improves visual acuity, and general physiological wellness.

- Artificial Lighting Systems

Indoor enclosures may necessitate artificial lighting systems to maintain constant photoperiods, imitate dawn and dusk transitions, and deliver UVB sunlight required for calcium metabolism and bone health in captive macaques. Full-spectrum lighting fixtures should be positioned to reduce glare, provide uniform illumination, and simulate natural lighting conditions.

Feeding stations and Dietary Management

Proper diet is critical for the health, growth, and lifespan of macaque monkeys in captivity. Enclosures should give access to balanced foods, nutritional supplements, and appropriate feeding stations that reduce competition while promoting individual eating behaviors.

- Nutritional requirements

Macaque diets should reflect their natural eating habits, including a variety of fruits, vegetables, leafy greens, protein sources, and extra vitamins or minerals as prescribed by veterinary nutritionists. Feeding schedules and portion control help to prevent obesity, dietary deficits, and digestive issues in caged monkeys.

- Feeding Station Design

Multiple feeding stations located throughout the cage eliminate competition and let all group members to obtain food at the same time. Elevated platforms, hanging feeders, and foraging gadgets enable macaques to feed, increase physical activity, and activate their natural foraging instincts.

Veterinary Care and Health Monitoring

Regular veterinarian care and health monitoring are critical components of competent macaque husbandry, allowing for early identification and treatment of medical issues, preventive healthcare interventions, and welfare assessments.

- Veterinary examinations

Comprehensive physical assessments, dental evaluations, immunizations, and parasite screens are all part of the routine veterinary examinations for macaques. Veterinarians who specialize in primate medicine provide tailored care regimens based on medical histories, behavioral observations, and test results.

- Health Monitoring Protocols

Ongoing health monitoring entails doing systematic examinations of bodily condition, reproductive status, behavioral changes, and fecal samples to detect indicators of gastrointestinal health, parasites, or infectious infections. Regular health records chronicle each macaque's illness history, treatment plan, and preventive healthcare practices.

Waste Management and Environmental Hygiene

Effective waste management procedures and environmental hygiene protocols reduce contamination risks, prevent disease transmission, and ensure clean, hygienic conditions in macaque enclosures.

- Waste Removal Systems

To reduce odor buildup and bacterial development, enclosure designs should include effective waste collection systems such drainable flooring, litter boxes, or substrate modifications. Cleaning routines and sanitation practices guarantee sanitary environments for macaque monkeys and caregivers.

Chapter 4

Diet and Nutrition for Macaque monkeys

Proper diet and nutrition are essential for the health, well-being, and lifespan of macaque monkeys in captivity. Macaques, like all primates, have certain nutritional needs that must be addressed in order to maintain their physiological processes, growth, and overall health. This detailed book delves into the dietary needs of macaque monkeys, including natural diet composition, nutritional requirements, feeding procedures, dietary enrichment, and prevalent health concerns.

Natural Diet Composition

Macaque monkeys are omnivorous primates with varying feeding habits and dietary preferences depending on their species, location, and ecological niche. Understanding macaques' natural diet composition provides useful information for constructing nutritionally appropriate diets in captivity.

- Wild Diet Variability

Macaques' natural diet varies according to their geographic location, habitat type, and seasonally available food supplies. Macaques typically eat a variety of fruits, seeds, nuts, leaves, flowers, insects, and tiny vertebrates, demonstrating their opportunistic feeding behavior and adaptive nutritional techniques.

- Species-Specific Preferences

Different macaque species may have unique nutritional preferences and foraging activities that are determined by their ecological environment. Rhesus macaques (Macaca mulatta) eat largely fruits, seeds, and invertebrates, but Japanese macaques (Macaca fuscata) consume a wider range of plant materials and coastal resources.

- Dietary adaptations

Macaques have evolved physiological adaptations to ingest and assimilate a diverse range of plant and animal meals. Their digestive systems are designed to absorb fibrous plant materials, extract nutrients from a variety of dietary sources, and deal with seasonal fluctuations in food availability.

Nutritional requirements

Meeting the dietary needs of macaque monkeys in captivity necessitates careful consideration of their species-specific requirements, life stage (e.g., juvenile, adult, old), reproductive status, and individual health concerns. A balanced diet promotes optimal growth, development, immunological function, and reproductive health in confined macaques.

- Macronutrients

Proteins are essential for muscle development, immunological function, and general growth. Macaques need high-quality protein sources such lean meats, eggs, dairy products, and plant-based proteins (e.g., legumes, nuts) to meet their amino acid needs.

Carbohydrates supply energy for daily activities and metabolic processes. Complex carbohydrates from fruits, vegetables, whole grains, and roots help macaques maintain intestinal health and energy levels.

Fats are concentrated energy sources that provide vital fatty acids (omega-3 and omega-6), which are required for brain function, coat condition, and reproductive health. Nuts, seeds, oils, and fatty fish are good sources (for marine species).

- Micronutrients

Vitamins are essential for a variety of physiological activities, including eyesight, immunological function, and tissue repair. Macaques require a well-balanced diet rich in vitamins A, B complex, C, D, E, and K, which they can acquire from fruits, leafy greens, vegetables, and fortified nutritional supplements.

Minerals: Essential for bone health, muscle function, and enzymatic activity. A macaque's diet must include calcium, phosphorus, magnesium, potassium, sodium, and trace minerals (such as zinc and iron), which can be obtained via leafy greens, fruits, nuts, and mineral supplements.

- Water

Proper hydration is critical to macaque health and metabolism. In captive macaques, fresh, clean water should always be supplied to prevent dehydration, assist digestion, maintain body temperature, and facilitate nutritional absorption.

- Feeding practices

Establishing appropriate feeding methods ensures that macaque monkeys receive balanced nutrition, maintain good eating habits, and reduce competition for food resources in group housing settings.

- Feeding Schedule

A constant feeding schedule suits macaques' normal foraging and digestion patterns. Regular meal times with several feeding sessions throughout the day reduce overfeeding, increase satiety, and foster natural eating behaviors in group-housed macaques.

- Portion Control

Monitoring portion sizes and food distribution helps confined macaques avoid obesity, nutritional imbalances, and digestive difficulties. Individualized feeding plans take into account age, gender, body condition, and activity levels to regulate calorie intake and guarantee appropriate nutrition without excessive weight gain.

- Food Enrichment

Food enrichment activities improve feeding experiences for macaque monkeys, stimulating both mental and physical engagement. Scatter feeding, puzzle feeders, foraging boxes, and food-dispensing devices are all enrichment tactics that improve problem-solving abilities, natural behaviors, and lessen feeding competition.

- Dietary Diversity

Offering a varied choice of foods mimics macaques' natural dietary variability and promotes nutritional balance in captivity.

Rotating fruits, vegetables, nuts, seeds, leafy greens, and protein sources (e.g., insects, eggs) reduces dietary monotony, improves palatability, and assures adequate nutritional intake.

- Dietary Enrichment

Dietary enrichment tactics are intended to improve the sensory, cognitive, and physical experiences connected with food consumption in confined macaques. These enrichments encourage natural foraging activities, mental stimulation, and social interactions in group housing environments.

- Sensory Enrichment

Introducing new food textures, aromas, and odors stimulates macaques' senses and fosters exploratory activity during feeding sessions. Scent trails, aromatic herbs, and edible flowers enhance the sensory experience of food consumption while also promoting dietary diversity.

- Cognitive Enrichment

Cognitive enrichment consists of food-related exercises that test macaques' problem-solving ability, memory retention, and learning capacities. Puzzle feeders, food puzzles, and interactive feeding systems demand macaques to manipulate things, use tactics, and participate in persistent food acquisition activities.

- Social Enrichment

Group feeding sessions and cooperative feeding activities help macaques develop social connections, communication skills, and cooperative behaviors in group housing conditions. Social enrichment through communal feeding experiences increases social relationships, decreases competitiveness, and improves group cohesion among troop members.

Common Health Considerations

Maintaining proper feeding practices and nutritional management helps to address frequent health issues in confined macaque populations. Regular veterinary exams, nutritional changes, and preventative healthcare measures help macaques maintain good health, wellness, and illness prevention.

- Dental Health

Proper nutrition and nutritional enrichment activities promote healthy chewing habits, reduce plaque formation, and prevent dental disease. Routine examinations and dental care routines allow for early detection and treatment of oral health disorders in macaque monkeys.

- Gastrointestinal Health

Captive macaques benefit from balanced diets high in fiber, water, and digestive enzymes, which promote gastrointestinal health and regular bowel movements. Gastrointestinal diseases such as constipation, diarrhea, and stomach disturbances are managed with dietary changes, probiotic supplements, and veterinary consultations.

- Metabolic disorders

Obesity and metabolic diseases in confined macaques might result from overfeeding, poor nutrition, or inactive lifestyles. Implementing portion management, food monitoring, and enrichment-based exercise programs can reduce obesity risks and improve metabolic health in macaques.

- Nutritional deficiencies

Inadequate nutrient intake or uneven diets can cause nutritional deficiencies in macaques, impacting their health and physiological activities. Monitoring food adequacy, consulting with veterinary nutritionists, and integrating vitamin or mineral supplements can help fill nutritional gaps and assure complete nutrient intake.

Diet and nutrition are critical in maintaining the health, wellbeing, and behavioral integrity of captive macaque monkeys. Caregivers and veterinary professionals can improve macaque health and quality of life in captive conditions by studying their natural food preferences, nutritional requirements, feeding techniques, and enrichment needs. Regular veterinary monitoring, food changes, and enrichment programs help to maintain nutritional wellbeing, prevent health issues, and increase overall well-being in macaque populations under human care.

Chapter 5

Health and veterinary care for macaque monkeys

Responsible macaque husbandry in captivity includes both health and veterinary care. Effective healthcare management includes preventive measures, medicinal interventions, welfare evaluations, and specialized veterinary procedures that are adapted to macaque monkeys' specific physiological, behavioral, and environmental demands. This thorough guide delves into major aspects of health and veterinary care, stressing proactive measures for promoting optimal health, diagnosing medical issues, and ensuring the well-being of captive macaque populations.

Preventive healthcare measures

Preventive healthcare treatments are critical for the overall health and well-being of macaques in captivity. These proactive tactics emphasize illness prevention, health promotion, and early diagnosis of medical issues through routine monitoring and preventative measures.

- Veterinary examinations

Scheduled veterinary checkups are critical for assessing health, recognizing early signs of illness, and implementing preventative healthcare measures in macaque populations. Comprehensive physical assessments, such as body condition scores, dental evaluations, and palpation of lymph nodes, organs, and the musculoskeletal system, provide information about overall health and detect anomalies.

- Vaccine Protocols

Vaccination procedures specific to macaque species and area disease risks defend against infectious illnesses found in both captive and wild populations. Veterinary advice and epidemiological considerations guide the administration of core vaccines such as tetanus toxoid and measles, as well as non-core vaccines related to local disease risks (e.g., tuberculosis, herpesvirus).

- Parasite Control Programs

Parasite management programs include monthly screenings, fecal examinations, and deworming treatments to prevent parasite infections that are common in confined macaques. Veterinary assessments and diagnostic findings inform the implementation of protocols for internal parasites (e.g., nematodes, cestodes) and external parasites (e.g., ectoparasites such as mites, ticks).

- Dental Care and Oral Health

Dental care is essential for preserving oral health, preventing dental problems, and boosting overall well-being in macaques. Routine dental checkups, preventative cleanings, and dental treatments (such as tooth extractions and root canal therapies) address dental anomalies, periodontal disease, and dental trauma caused by captive situations.

- Diagnostic Procedures

Diagnostic methods are useful for evaluating medical problems, identifying illnesses, and determining treatment regimens for macaque monkeys in veterinary care. These treatments use advanced diagnostic tools, methodologies, and laboratory studies to assess health and aid in clinical decision-making.

- Blood and Serum Analysis

Blood tests, such as complete blood counts (CBC), serum chemistry panels, and serological assays (e.g., ELISA, PCR), can offer useful information about macaque hematological parameters, biochemical profiles, and immunological responses. These diagnostic tests identify systemic ailments, infectious infections, and metabolic imbalances that necessitate medical care.

- Imaging studies

Imaging studies, such as radiography (X-rays), ultrasonography, and computed tomography (CT scans), allow for non-invasive evaluations of anatomical features, organ function, and pathological alterations in macaque monkeys. Diagnostic imaging helps to diagnose musculoskeletal injuries, abdominal abnormalities, and reproductive diseases in captive macaques.

- Fecal and Urinary Analysis

In macaque populations, fecal and urine study includes microscopic examination, culture, and biochemical assays to assess gastrointestinal health, renal function, and metabolic processes. These diagnostic tests identify gastrointestinal parasites, urinary tract infections, and metabolic problems that necessitate dietary changes or therapeutic therapies.

- Medical interventions

Medical interventions include therapeutic treatments, surgical operations, and pharmaceutical therapies provided by veterinary specialists to manage medical disorders, ease symptoms, and improve recovery in macaques.

- Pharmaceutical therapies

Antibiotics, antiparasitics, analgesics, and anti-inflammatory drugs are used to treat infectious infections in macaques, as well as to manage pain and inflammation. Individualized treatment regimens take into account species-specific drug sensitivity, dosage calculations, and therapeutic monitoring to guarantee efficacy and safety.

- Surgical procedures

Sterilization surgeries (e.g., ovariohysterectomy, castration), dental surgeries, and exploratory surgeries are all used to treat reproductive issues, dental abnormalities, and internal injuries in macaques. Veterinary surgical protocols focus on anesthesia safety, aseptic practices, and post-operative care in order to improve surgical outcomes and reduce problems.

- Wound Care and Trauma Management

Wound treatment methods for macaque monkeys include cleansing, debridement, and dressing of traumatic injuries, lacerations, or surgical wounds. Veterinary wound management solutions use topical treatments, bandaging techniques, and systemic drugs as needed to enhance tissue healing, prevent infection, and ease suffering.

Behavioral Health and Welfare

Behavioral health and welfare factors are critical for enhancing psychological well-being, lowering stress, and maintaining natural behaviors in macaque monkeys in captivity. Behavioral assessments, enrichment programs, and social management measures help to improve quality of life and reduce behavioral anomalies.

- Behavioral Assessments

Behavioral assessments measure social interactions, activity patterns, and responses to environmental stimuli in macaque populations. Observational studies, behavioral scoring systems (e.g., ethograms), and behavioral tests (e.g., cognitive tasks) all contribute to enrichment methods, social housing dynamics, and personalized behavioral management plans.

- Environmental Enrichment

Captive macaque monkeys benefit from environmental enrichment programs, which give cognitive stimulation, physical exercise, and social interaction chances. Foraging puzzles, sensory stimuli (e.g., scent trails), and social grooming sessions are examples of enrichment activities that promote natural behaviors, minimize stereotyping, and improve overall welfare.

- Social Management

Social management tactics seek to improve group dynamics, reduce aggression, and promote positive social interactions among macaque troop members in group housing environments. Behavioral observations, social compatibility assessments, and progressive introductions guide group composition decisions and foster harmonious social structures.

Geriatric Care and Age-Related Health Concerns

Geriatric care addresses age-related health difficulties, cognitive changes, and chronic medical diseases that afflict senior macaque monkeys in captivity. Specialized veterinary care, geriatric examinations, and supportive interventions improve quality of life, manage age-related illnesses, and provide compassionate end-of-life care when needed.

- Geriatric Health Assessments

Geriatric health assessments track physiological changes, mobility constraints, and cognitive deterioration associated with aging in macaques. Routine geriatric screenings, mobility evaluations, and cognitive function tests detect age-related health issues and inform supportive care plans for senior macaques.

- Pain Management and Palliative Care

Pain treatment guidelines prioritize pain assessment, analgesic medications, and palliative care approaches in geriatric macaque monkeys to reduce discomfort and increase quality of life. Chronic pain disorders and age-related mobility impairments are addressed using multimodal pain treatment techniques that include pharmaceutical interventions and supportive therapies (for example, physical therapy and acupuncture).

Ethical Concerns and Welfare Guidelines

Ethical considerations and welfare guidelines emphasize the relevance of ethical norms, humane treatment, and species-specific welfare procedures when handling macaque monkeys in captivity. Adherence to ethical standards, legal requirements, and best practices ensures that macaque research and conservation initiatives are conducted responsibly, with an emphasis on animal welfare and scientific integrity.

- Ethical Standards for Research

Ethical standards in macaque research prioritize animal welfare, reduce experimental hazards, and maintain scientific rigor in compliance with institutional animal care norms (for example, IACUC protocols). Ethical review boards, research oversight committees, and animal welfare assessments help to assure ethical compliance, promote transparency, and advocate for macaque wellbeing in research settings.

- Regulatory Compliance

Compliance with local, national, and international animal welfare rules and regulations dictates the ethical treatment, living circumstances, and veterinary care methods for captive macaque monkeys. Compliance monitoring, regulatory inspections, and accreditation processes ensure that legal standards are met, accountability is promoted, and individual welfare is protected across a wide range of organizational contexts.

Health and veterinary care are critical components of good macaque husbandry, and include preventative healthcare measures, diagnostic techniques, medicinal interventions, behavioral health considerations, and ethical guidelines specific to captive situations. Caregivers and veterinary experts improve the health, well-being, and longevity of macaque monkeys under human care by focusing on proactive healthcare management, supporting welfare-centric methods, and incorporating specialist veterinary care. Continuous advances in veterinary medicine, welfare research, and behavioral enrichment help to raise care standards and advocate for better health results in macaque populations worldwide.

Chapter 6

Enrichment and mental stimulation for macaque monkeys

Enrichment and mental stimulation are essential components of animal care for macaque monkeys, ensuring their psychological well-being and encouraging natural behaviors in captivity. Enrichment activities are intended to imitate the intricacies of their natural environments, allowing for physical activity, cognitive challenges, and social interactions. This detailed book goes into the importance of enrichment, the various types of enrichment, implementation tactics, and the benefits to captive macaques' health and welfare.

Importance of Enrichment

Enrichment is critical for improving the general health of macaque monkeys in captivity. It meets the physical, psychological, and social demands of these clever and highly social primates, minimizing boredom, lowering stress, and improving their quality of living.

- Psychological Well Being

Enrichment activities boost macaques' cognitive capabilities, reducing the formation of stereotypic behaviors that are commonly associated with boredom and mental stress. Engaging their minds with complex tasks, puzzles, and new experiences improves their mental and emotional wellness.

- Physical Health

Physical enrichment helps macaques to participate in natural behaviors like climbing, foraging, and exploring, which are important for their physical fitness and general health. Active participation in enrichment activities improves muscle growth, coordination, and cardiovascular health.

- Social Interactions

Macaques are naturally gregarious animals who flourish in groups. Enrichment that encourages social connections, such as cooperative tasks and grooming chances, develops social relationships, lowers aggression, and promotes peaceful group dynamics.

Types of Enrichment

Effective enrichment programs include a wide range of activities that address many areas of macaque behavior and natural proclivities. Physical, cognitive, sensory, social, and food enrichment are the five basic forms.

- Physical Enrichment

Physical enrichment is creating environments and activities that promote physical activity, exploration, and play. This form of enrichment is critical for the physical health and agility of macaques.

- Climbing Structures: Providing ropes, ladders, branches, and platforms encourages climbing and exploring, which mimics the arboreal behaviors that macaques participate in in the wild.
- Obstacle Courses: Designing obstacle courses with varied degrees of difficulty requires macaques to traverse, climb, and jump, increasing physical activity and problem solving.
- Introducing objects like balls, swings, and hanging toys encourages play behavior, which improves physical fitness and mental engagement.

- Cognitive Enrichment

Cognitive enrichment tries to test macaques' mental talents through problem-solving exercises, riddles, and educational activities. This form of enrichment is essential for avoiding cognitive decline and increasing mental stimulation.

- Puzzle Feeders: Using puzzle feeders that involve manipulation to acquire food improves cognitive processes and problem-solving abilities.
- Positive reinforcement training sessions teach macaques new actions and orders, giving mental challenges while also improving the link between animals and humans.
- Novelty stuff: Placing new objects, toys, or strange stuff in the enclosure stimulates curiosity and fosters exploratory activity, hence increasing cognitive engagement.

- Sensory Enrichment

Sensory enrichment activates macaques' many senses, improving their sensory awareness and providing unique experiences that replicate their natural habitat.

- Scent Trails: Making scent trails with herbs, spices, or natural scents promotes sensory exploration and foraging activity.
- Auditory Stimuli: Playing recordings of natural sounds, such as bird calls, water streams, or rainforest ambiance, gives auditory stimulation and an appreciation for environmental diversity.
- Tactile Enrichment: Providing a range of textures, such as leaf, bark, or substrates, stimulates the tactile senses and encourages natural exploration.

- Social Enrichment

Social enrichment aims to improve good social interactions among group-housed macaques by encouraging social connections, collaboration, and natural group dynamics.

- Grooming opportunity: Providing grooming equipment or opportunity for macaques to groom one another promotes social bonding and decreases stress.
- Cooperative Tasks: Creating tasks that need cooperation, such as group puzzle feeders or shared enrichment activities, promotes collaboration and social cohesion.
- Social Introductions: Carefully planned introductions of new group members, combined with slow and monitored interactions, improve social dynamics and group stability.

- Dietary Enrichment

Dietary enrichment is improving the feeding experience by providing variety, complexity, and foraging opportunities that replicate natural feeding behaviors.

- Foraging Activities: Scattering food, hiding goodies, or utilizing foraging devices encourages natural foraging behaviors and increases feeding time.
- Dietary Variety: Providing a broad selection of fruits, vegetables, nuts, and protein sources helps to avoid dietary monotony and improves nutritional balance.
- Food Puzzles: Using food puzzles that involve manipulation to acquire food promotes cognitive engagement while simulating natural foraging problems.

Implementation Strategies

Enrichment programs must be carefully planned, assessed on a regular basis, and adapted to meet the macaque population's individual needs and behaviors. Environmental design, individual enrichment plans, and caregiver involvement are all important techniques for delivering effective enrichment.

- Environmental Design

Designing an enriching environment entails developing a dynamic and exciting habitat that meets the physical, cognitive, and social demands of macaques.

- Spatial Complexity: Providing a range of spatial aspects, such as varied levels, hiding places, and open regions, promotes exploration and natural behaviours.

- Rotating Enrichment: Keeping enrichment objects and activities fresh minimizes boredom and keeps macaques engaged and mentally challenged.
- Natural components: Including natural components like real plants, natural soils, and water features increases environmental complexity and mimics natural environments.

Individualized Enrichment Plans

Individualized enrichment plans take into account each macaque's specific preferences, needs, and behaviors, ensuring that enrichment activities are tailored to optimize effectiveness.

- Behavioral Observations: Regular behavioral observations assist uncover individual preferences, interests, and indicators of stress or boredom, which shape individualized enrichment initiatives.
- Age and Health Considerations: Tailoring enrichment activities to macaques' ages, health statuses, and physical capacities ensures that everyone benefits from the program.
- Preference Assessments: Periodically assessing macaques' preferences and responses to various enrichment activities helps guide the development and implementation of effective enrichment initiatives.

Caregiver Involvement

Caregivers' active participation is critical to the effectiveness of enrichment programs since they are responsible for implementing activities, assessing responses, and changing techniques based on continuing assessments.

- Training and Education: Teaching caregivers about enrichment principles, practices, and the necessity of mental stimulation improves their ability to execute effective programs.
- Daily Interaction: Through positive reinforcement training and enrichment activities, caregivers and macaques can develop their bond and increase their general well-being.
- Feedback and Adaptation: Encouraging caregivers to submit feedback on the success of enrichment activities and making required changes ensures that the program remains dynamic and sensitive to macaques' needs.

The Effects of Enrichment on Health and Welfare

Implementing thorough enrichment programs has a significant positive influence on the health, welfare, and overall quality of life of caged macaques. These effects are visible in different aspects of their behavior, physiology, and social interactions.

Behavioral Benefits
Enrichment activities greatly improve macaques' behavioral repertoire by promoting natural behaviors, lowering stereotypic acts, and increasing overall activity levels.

- Engaging in enrichment activities reduces the prevalence of stereotypic behaviors such as pacing, self-injury, and repetitive motions, indicating better mental health.

- Increased Activity Levels: Enrichment promotes macaques to be more active, investigate their surroundings, and play, resulting in improved physical fitness and health.
- Enhanced Cognitive Function: Cognitive enrichment exercises test macaques' problem-solving ability and mental sharpness, preventing cognitive decline and promoting lifelong learning.

Physiological Benefits

The physiological benefits of enrichment are shown in greater physical health, lower stress levels, and improved immune function in macaque populations.

- Improved Physical Health: Physical enrichment activities encourage exercise, muscle growth, and cardiovascular health, lowering the risk of obesity and other health problems.
- Reduced Stress Levels: Enrichment activities that give mental stimulation and social contact lower stress hormones like cortisol while also promoting relaxation and well-being.
- Enhanced immunological Function: Macaques that are engaged and mentally stimulated often have greater immunological responses, which leads to improved overall health and disease resistance.

Social Benefits:

Enrichment that promotes positive social interactions helps group-housed macaques have more harmonious group dynamics, stronger social ties, and better welfare.

- Social enrichment activities, such as cooperative chores and grooming chances, help to increase social cohesion and deepen bonds among group members.
- Reduced Aggression: Structured social interactions and enrichment diminish aggressive behavior and dominance disputes, resulting in a more stable and peaceable social environment.
- Improved Group Dynamics: Enrichment that promotes natural social behaviors and hierarchical structures improves group dynamics, lowering stress and increasing social harmony.

Case Studies & Examples

The beneficial effects of enrichment on macaque monkeys have been documented in several case studies and examples from zoos, research facilities, and sanctuary contexts.

- Case Study 1: Puzzle Feeders at a Zoo

In a zoo context, the addition of puzzle feeders to a group of rhesus macaques resulted in a dramatic reduction in stereotypic behaviors and greater involvement in problem-solving tasks. During feeding sessions, the macaques showed better cognitive ability, increased physical activity, and improved social connections. Caregivers noticed that the macaques engaged in more natural foraging behaviors and spent more time investigating their surroundings, indicating greater overall well-being.

- Case Study 2: Sensory Enrichment at a Research Facility

A study facility established a sensory enrichment program for a group of Japanese macaques, which included scent trails, aural stimuli, and tactile objects. The macaques showed enhanced exploratory activity, heightened interest, and lower stress levels. The enrichment program also helped macaques form stronger social bonds, as they frequently engaged in cooperative exploration and shared sensory experiences. Researchers discovered that sensory enrichment helped to improve physiological health, such as decreased stress hormone levels and better immunological function.

- Case Study 3: Social Enrichment in a Sanctuary Environment

A social enrichment program was developed in a sanctuary setting to benefit a group of long-tailed macaques. Cooperative tasks, grooming opportunities, and well planned social introductions were used to encourage favorable social connections. The macaques demonstrated stronger social ties, less aggression, and better group dynamics. Caregivers observed that the macaques appeared more relaxed, showed fewer indications of stress, and engaged in normal social behaviors, indicating a higher quality of life.

Enrichment and mental stimulation are critical components of macaque husbandry, maintaining the psychological well-being, physical health, and social harmony of confined macaques. Comprehensive enrichment programs that include physical, cognitive, sensory, social, and dietary activities encourage natural behaviors, reduce boredom, and improve overall well-being. Caregivers and veterinary professionals can dramatically improve macaque monkeys' quality of life by applying well-designed enrichment measures, resulting in a more stimulating and fulfilling environment for these clever and social animals. Continuous assessment, modification, and innovation in enrichment methods help to advance care standards and advocate for the welfare of captive macaques.

Chapter 7

Training and socialization

Training and socialization are important aspects of macaque monkey care and management in captivity. Effective training methods improve macaque welfare by encouraging positive behaviors, improving veterinarian treatment, and enhancing their lives. Socialization, on the other hand, ensures that macaques form and maintain strong social ties, which reduces stress and promotes general health. This comprehensive reference discusses the principles, methods, benefits, problems, and practical applications of captive macaque training and socialization.

Importance of Training

Training is essential for managing and caring for macaque monkeys. It not only improves the quality of care, but it also enhances the lives of these clever primates by offering mental stimulation and opportunity for pleasant relationships with caretakers.

- **Enhancing Welfare**
Positive reinforcement-based training regimens can greatly improve macaque welfare. Training helps to create a more positive and enriching environment by encouraging desired behaviors while lowering stress.

- Mental Stimulation: Training provides cognitive challenges that keep macaques mentally engaged and stimulated, minimizing boredom and lowering the risk of stereotypic behaviour.
- tension Reduction: Positive reinforcement training fosters trust between macaques and their caregivers, lowering anxiety and tension during routine husbandry and veterinary operations.
- Empowerment: Training gives macaques choices and control over their surroundings, which can improve their overall health and minimize stress.

- **facilitating veterinary care**

Training macaques to willingly participate in veterinary operations can substantially improve their care and management. Cooperative training lowers the need for physical restraint and anesthesia, making medical treatments safer and less traumatic for both animals and their caregivers.

- Voluntary Participation: Teaching macaques to enter transport crates, present body parts for examination, and participate with injections or blood draws can improve veterinary operations and reduce stress.
- Routine Health Monitoring: Caregivers can regularly monitor macaques' health and behavior through regular training sessions, allowing for early detection of medical disorders and prompt management.
- Minimizing Restraint: Training lowers the need for physical restraint, which can be stressful and sometimes damaging,

improving macaques' general well-being during medical operations.

- **Enrichment and Behavioral Health**

Training is a type of enrichment that encourages natural behaviors while preventing boredom. Training improves macaques' behavioral health and quality of life by engaging them in problem-solving exercises and interactive tasks.

- Natural Behaviors: Training regimens that encourage foraging, climbing, and social interactions promote natural behaviors, which are critical to macaques' psychological well-being.
- Behavioral Diversity: Training incorporates novel and different activities into macaques' everyday routines, boosting behavioral diversity while lowering the danger of stereotypic behavior.
- pleasant Interactions: Training sessions encourage pleasant interactions between macaques and caretakers, which strengthens the human-animal link and fosters trust.

Principles of Training

Effective training methods for macaque monkeys are founded on scientific concepts of behavior modification, notably positive reinforcement. These principles ensure that animals receive training that is humane, effective, and fun.

- **Positive Reinforcement**

Positive reinforcement is the process of rewarding desired behaviors in order to improve the possibility that they will be repeated. This strategy works especially well with macaques, who are highly driven by food, social contacts, and other rewards.

- Rewards may include favored foods, accolades, toys, or access to chosen activities. The objective is to provide rewards that are meaningful and motivating to the individual macaque.
- Timing: Reinforcers must be administered shortly following the intended behavior to ensure that the macaque associates the behavior with the reward.
- Consistency: The consistent use of positive reinforcement aids in the establishment of clear links between activities and rewards, hence improving learning.

- **Shaping**

Shaping entails encouraging successive approximations to a desired behavior. This strategy is particularly beneficial for breaking down complex activities into smaller, more manageable phases.

- Incremental Steps: The desired behavior is divided into small, manageable steps, with each step reinforced until the entire behavior is acquired.
- Gradual Progression: As the macaque advances through the training steps, trainers gradually increase the criteria for reward, ensuring that the learning process runs smoothly.

- Patience and Flexibility: Shaping necessitates patience and flexibility because macaques progress at different speeds and may require additional assistance at various phases.

- **Desensitization and Counterconditioning**

Desensitization and counterconditioning procedures are used to diminish fear and anxiety caused by specific stimuli or situations. These techniques are especially beneficial for preparing macaques for veterinary operations and other potentially stressful situations.

- Gradual Exposure: Desensitization entails exposing the macaque to a feared stimulus at a low intensity and gradually increasing the intensity over time.
- Counterconditioning involves associating the scary stimuli with good experiences, such as food or praise, in order to alter the macaque's emotional response.
- Building Confidence: These strategies assist macaques gain confidence and minimize anxiety, making them more comfortable and cooperative in a variety of settings.

Training Techniques and Methods

Macaques can be taught certain actions and tasks using a variety of training approaches and methodologies. The technique used is determined by the specific animal, intended behavior, and training goals.

- **Clicker Training**

Clicker training employs a small, handheld device that emits a clicking sound to indicate the precise instant a desired behavior happens. The click is followed by a reward, which reinforces the behavior.

- Marking action: The clicker acts as a clear and constant marker, informing the macaque that they have completed the appropriate action.
- Precision: Clicker training promotes exact timing and communication, making it easier to mold complicated actions.
- happy Associations: The clicker becomes associated with happy experiences, which improves training effectiveness.

- **Target Training**

Target training teaches macaques to touch or follow a specific object, such as a stick or a hand. This strategy is excellent for guiding macaques to specific areas or positions.

- Introducing the Target: The macaque is shown the target and is rewarded if it touches or follows it.
- Building on the Behavior: Once the macaque consistently touches or follows the target, the behavior can be expanded to direct them to new areas or do specific activities.
- Versatility: Target training is a versatile strategy that may be applied to a variety of situations, including husbandry, veterinary treatment, and enrichment.

- **Station Training**

Station training teaches macaques to move to and stay in a certain region or "station" on cue. This strategy is effective for regulating group dynamics and facilitating husbandry duties.

- Establishing the Station: A specific station, such as a mat or a perch, is introduced, and the macaque is rewarded for moving to and remaining on it.
- Cueing the Behavior: The macaque is trained to move to the station when prompted, and this behavior is rewarded with rewards.
- Managing Group Dynamics: Station training facilitates group interactions by giving a clear and predictable structure, hence minimizing conflicts and stress.

Importance of Socialization

Socialization is critical for the psychological well-being of macaque monkeys. It refers to the process of establishing and maintaining good social ties with conspecifics and, in certain situations, humans.

- **Natural social behaviors**
Macaques are extremely gregarious animals who engage in a variety of intricate social behaviors. Proper socialization ensures that these behaviors are expressed in a healthy and appropriate manner.

- Grooming is an important social action that promotes social relationships and minimizes tensions within groups.

- Play behavior is essential for the development of social skills, cognitive capacities, and physical coordination in juvenile macaques.
- Socialization contributes to the formation and maintenance of stable hierarchical structures, which are necessary for group cohesion and stability.

- **Reduced Stress and Aggression**

Proper socialization minimizes stress and violence by encouraging pleasant social interactions and minimizing intragroup conflicts.

- Positive Interactions: Socialization promotes positive interactions like grooming and cooperation, which alleviate stress and improve group cohesion.
- Conflict Resolution: Macaques who are socially well-adjusted are better able to resolve problems amicably, which reduces the frequency of aggressive behavior.
- Stress Reduction: Studies have shown that social interactions, particularly grooming, can lower stress hormones and increase relaxation and well-being.

Socialization Strategies

Effective socialization tactics are critical for developing appropriate social behaviors and ensuring the psychological well-being of macaques. These tactics include carefully planned introductions, social groups, and ongoing monitoring.

- **Gradual introductions**

Gradual introductions are critical for ensuring that new members integrate successfully into existing groups while reducing stress and violence.

- Observation and Assessment: Prior to introductions, individual behaviors, temperaments, and social dynamics are carefully observed and assessed to determine possible compatibility.
- Visual and olfactory interaction: New members are initially able to see and smell one other without making direct physical touch, which reduces stress and promotes familiarity.
- Controlled Introductions: Gradual, controlled introductions with close supervision allow macaques to socialize securely, with prompt intervention if conflicts emerge.

- **Social Grouping**

Social grouping tactics entail building stable, compatible groupings that encourage pleasant social interactions and reduce conflict.

- Compatibility evaluations: Conducting regular compatibility assessments helps identify and address social conflicts, ensuring that groups stay stable and harmonious.
- Balanced Group Composition: Groups are made up of people with similar temperaments, ages, and social roles, resulting in balanced social dynamics.

- Monitoring and Adjustment: Constant monitoring and adjustment of group compositions helps to reduce social tensions and boost pleasant interactions.

Challenges and Solutions for Training and Socialization

Individual variances, environmental limits, and controlling group dynamics all contribute to the difficulties of training and socializing macaques. Effective solutions and strategies are required to overcome these problems and achieve effective outcomes.

- **Individual Differences**

Individual temperament, learning abilities, and social behaviors vary among macaques, which can have an impact on training and socialization efforts.

- Tailored Approaches: Training and socialization programs must be tailored to each macaque's unique needs and preferences, ensuring that activities are engaging and appropriate.
- Patience and flexibility are essential for adapting to individual differences and supporting each macaque's progress.
- Positive Reinforcement: The consistent use of positive reinforcement promotes trust and confidence, facilitating successful training and socialization.

- **Environmental constraints**

Environmental constraints, such as limited space and resources, can impede effective training and socialization programs.

- Innovative and creative solutions, such as using portable training equipment and making the best use of available space, can help overcome environmental constraints.
- Enrichment Integration: By incorporating training and socialization activities into daily enrichment routines, macaques receive consistent and meaningful stimulation.
- Resource Allocation: Effective resource allocation, such as time, materials, and caregiver support, ensures the long-term viability and effectiveness of training and socialization programs.

- **Managing group dynamics**

Macaque monkey socialization presents ongoing challenges in terms of managing group dynamics and addressing social tensions.

- Proactive Monitoring: By monitoring social interactions and intervening early in conflicts, group dynamics can be kept stable and harmonious.
- Conflict Resolution Strategies: Implementing conflict resolution strategies, such as providing escape routes and expanding enrichment opportunities, lowers the risk of aggression and stress.
- Supportive Environment: Providing adequate space, resources, and social opportunities promotes positive social interactions and reduces tension.

Case Studies & Examples

Numerous case studies and examples from zoos, research facilities, and sanctuaries demonstrate the successful implementation of training and socialization programs.

Case Study 1: Cooperative Veterinary Care in a Zoo Setting
A zoo implemented a cooperative veterinary care program for a group of long-tailed macaques, using positive reinforcement training to encourage voluntary participation in medical procedures. The macaques were trained to present body parts for examination, enter transport boxes, and comply with injections. The program significantly reduced stress during veterinary procedures, improved the macaques' overall well-being, and strengthened the bond between animals and caregivers.

Case Study 2: Social Enrichment at a Research Facility
A research facility introduced a social enrichment program for a group of rhesus macaques, focusing on grooming opportunities, cooperative tasks, and structured play sessions. The macaques exhibited increased social cohesion, reduced aggression, and enhanced overall welfare. The program also facilitated better group dynamics, promoting a more stable and harmonious social environment.

Case Study 3: Training for Enrichment in Sanctuary
A sanctuary developed a comprehensive training and enrichment program for a group of Japanese macaques, incorporating clicker training, target training, and cognitive puzzles. The macaques showed significant improvements in cognitive function, behavioral

diversity, and physical fitness. The training program also provided valuable enrichment, keeping the macaques mentally stimulated and engaged.

Training and socialization are essential components of the care and management of macaque monkeys in captivity. Effective training programs, based on positive reinforcement and other scientific principles, enhance the welfare of macaques by promoting mental stimulation, facilitating veterinary care, and enriching their lives. Proper socialization helps that macaques create and maintain healthy social ties, lowering stress and boosting general well-being. By employing well-designed training and socialization tactics, caregivers and veterinary professionals may dramatically improve the quality of life for macaque monkeys, enabling a more stimulating and fulfilling environment for these clever and social animals. Continuous assessment, adaptability, and innovation in training and socialization procedures help to raising standards of care and advocating for the well-being of macaques in captivity.

Chapter 8

Potential Challenges and Risks of Keeping Macaque Monkeys as Pets

Macaque monkeys, with their intelligence, social nature, and engaging behaviors, can seem like intriguing pets. However, keeping these primates as pets comes with significant challenges and risks that potential owners must carefully consider. These challenges range from legal and ethical concerns to health risks and behavioral issues. Understanding these factors is crucial for ensuring the welfare of the animals and the safety of their human caretakers.

Legal considerations

- Regulations and Permits

The legality of keeping macaque monkeys as pets varies widely across different jurisdictions. In many places, it is illegal or highly regulated due to the complexities and potential dangers involved.

State and Local Laws: Many states and municipalities have specific laws regarding the ownership of exotic animals, including macaque monkeys. Prospective owners must research and comply with these regulations to avoid legal repercussions.

Permits and Licensing: In areas where ownership is allowed, obtaining the necessary permits and licenses is often a rigorous process. This may include inspections, proof of adequate facilities, and adherence to strict care standards.

International Laws: For those considering importing macaques from other countries, international laws such as CITES (Convention on International Trade in Endangered Species) must be considered. Violation of these regulations may result in serious fines.

- Legal risks

Owning a macaque monkey can expose individuals to significant legal risks, including liability for injuries and property damage.

Personal Injury Liability: If a macaque injures someone, the owner can be held legally responsible. This can result in lawsuits, financial penalties, and even the forced removal of the animal.

Property Damage: Macaques are strong and curious animals that can cause significant damage to property. Owners must be prepared for the financial burden of repairs and replacements.

Zoning and Housing Restrictions: Many residential areas have zoning laws that prohibit the keeping of exotic animals. Violating these laws can lead to fines and eviction.

Health Risks

- Zoonotic diseases

Macaques can carry diseases that are transmissible to humans, known as zoonotic diseases. These diseases pose serious health risks to both the owner and the broader community.

Herpes B Virus: One of the most serious zoonotic risks associated with macaques is the Herpes B virus (Herpes simian B), which can be fatal to humans. Transmission can occur through bites, scratches, or contact with bodily fluids.

Tuberculosis: Macaques can carry tuberculosis, a highly contagious respiratory disease that can be transmitted to humans through close contact.
Simian Immunodeficiency Virus (SIV): Similar to HIV, SIV can be transmitted to humans, posing significant health risks.

Other Zoonoses: Additional diseases, such as hepatitis, rabies, and different parasite infections, can also be transmitted from macaques to people.

- Allergies and Respiratory Issues
Macaques can create allergens that may cause respiratory difficulties or allergic reactions in sensitive persons.

Dander and Fur Allergens: Similar to household pets, macaques create dander and shed fur, which can aggravate allergies.

Respiratory Irritants: Dust from bedding, urine, and feces can exacerbate respiratory diseases like asthma in people.

Behavioral Challenges

- Aggression and Territoriality

Macaques are naturally territorial and can display aggressive behaviors, especially during mating seasons or when they feel threatened.

Biting and Scratching: Macaques have sharp teeth and nails, and they can deliver severe bites and scratches. This aggression can be unpredictable and dangerous.

Dominance Hierarchies: In the wild, macaques establish complex social hierarchies. In captivity, they may attempt to assert dominance over their human caretakers, leading to conflicts and aggressive encounters.

Hormonal Changes: Males, in particular, can become highly aggressive during mating seasons, posing additional risks to owners.

- Social Needs and Isolation

Macaques are highly social animals that require extensive social interaction. Keeping a single macaque as a pet can lead to severe behavioral issues.

Social Deprivation: Without the companionship of other macaques, these primates can experience severe stress, anxiety, and depression.

Behavioral Problems: Isolated macaques may develop abnormal behaviors, such as self-mutilation, pacing, and repetitive motions, which are signs of psychological distress.

Enrichment demands: Meeting macaques' complex social and environmental enrichment demands in a family setting is highly difficult and frequently inadequate.

Environmental Challenges

- Space requirements

Macaques require expansive, complex settings that allow for climbing, foraging, and social interactions. Typical domestic settings are frequently insufficient.

Enclosure Size: Macaques require spacious enclosures that allow them vertical movement. Small cages or limited living spaces can cause physical and psychological problems.

Environmental Enrichment: Creating a stimulating environment with toys, puzzles, and various terrain is critical for reducing boredom and encouraging natural behaviors.

Outdoor Access: Regular access to outdoor environments is essential for macaques, as it allows for natural sunlight exposure and more diverse enrichment opportunities.

- Diet & Nutrition

Meeting the dietary needs of macaques is complex and requires careful planning and resources.

Nutritional Requirements: Macaques have specific dietary needs that include a variety of fruits, vegetables, insects, and specially formulated primate diets. Inadequate nutrition can lead to health problems such as obesity, malnutrition, and metabolic bone disease.

Foraging Opportunities: In the wild, macaques spend a lot of time foraging. To replicate this natural behavior in captivity, creative and compelling feeding opportunities must be provided.

Food Costs: Providing a balanced diet for macaques can be expensive and requires a steady supply of fresh, high-quality ingredients.
Ethical considerations

- Welfare and Quality of Life

Keeping macaques as pets raises significant ethical concerns regarding their welfare and quality of life.

Captivity Stress: The stress of captivity can have profound effects on macaques, leading to physical and psychological suffering.

Ethical Treatment: Ensuring that macaques receive ethical and humane treatment involves providing for their complex needs, which is often beyond the capabilities of individual pet owners.

Lifespan and Commitment: Macaques have long lifespans, frequently surviving 25 to 30 years in captivity. This long-term commitment can be tough for owners, leading to circumstances where the animals are abandoned or committed to sanctuaries.

- Conservation and Breeding

The pet trade can significantly affect wild macaque populations and conservation efforts.

illicit Wildlife Trade: The desire for pet macaques can feed illicit wildlife trade, leading to the capture and exploitation of natural populations.

Conservation Impact: Removing individuals from wild groups can destabilize social systems and undermine conservation efforts.

Breeding in Captivity: Breeding macaques for the pet trade can add to overpopulation and the ethical quandary of breeding creatures who may suffer in captivity.

Practical challenges

- Care and Maintenance

The everyday care and upkeep of macaques need tremendous time, effort, and resources.

Daily Routine: Caring for macaques involves a demanding daily routine, including feeding, cleaning, social engagement, and enrichment activities.

Veterinary Care: Finding veterinarians with expertise in primate care can be tough and expensive. Macaques require regular health check-ups, immunizations, and emergency care to stay healthy.

Long-Term Commitment: Owners must be prepared to make long-term commitments to care for macaques, which can last several decades. This commitment encompasses financial, emotional, and physical investments.

- Public Perception and Social Stigma

Keeping macaques as pets may result in public attention and social censure.

Neighbors and community members may voice worries about the safety and wellbeing of macaques in a residential environment.

Owners of unusual pets, such as macaques, may face social isolation as a result of their pets' unconventional nature and the possibility of negative public impression.

Owners may need to engage in lobbying and education activities to clear up myths and promote responsible pet ownership.

- Case Studies & Examples

Case Study 1: Behavioral issues in a captive macaque

A family adopted a newborn macaque monkey without fully comprehending the nuances of its upbringing. As the macaque aged, it began to exhibit aggressive tendencies, including as biting and scratching family members. The family struggled to provide appropriate social connection and environmental enrichment for the macaque, resulting in severe stress and behavioral difficulties. Eventually, the monkey had to be donated to a sanctuary, exposing the difficulties and dangers of keeping macaques as pets.

Case Study 2: Zoonotic Disease Outbreak

Another incidence involved a pet macaque transmitting the Herpes B virus to its owner via a bite. The owner experienced serious neurological problems and required extensive medical care. This instance highlights the serious health hazards posed by zoonotic infections, as well as the significance of safe handling and veterinary care.

Case Study 3: Ethical and Legal Implications.

A person in a location where keeping macaques as pets is forbidden was charged with a crime after neighbors reported the exotic animal's existence. The authorities confiscated the macaque, and the owner was fined and sentenced. This case demonstrates the legal and ethical issues of owning macaques, as well as the potential penalties of breaching legislation.

Keeping macaque monkeys as pets has a number of obstacles and concerns that go beyond the standard duties of pet keeping. These difficulties have legal, physiological, behavioral, environmental, ethical, and practical components. Before selecting to keep a macaque as a pet, potential owners should carefully evaluate these considerations and assess their capacity to meet their animal's complicated needs. Finally, the wellbeing of the animals and the safety of their owners should be the key factors when making responsible and ethical judgments about owning these intelligent and social primates. Ensuring that macaques receive adequate care, enrichment, and social interaction is critical for their well-being, and this is frequently best accomplished in professional facilities rather than private households.

Chapter 9

FAQs: Keeping Macaque Monkeys as Pets

1. Is it allowed to have macaque monkeys as pets?
The legality of having a macaque monkey as a pet varies widely depending on where you live. Many governments and countries make it unlawful to keep macaques because to the complexity and risks associated with their care.

- United States: Many states have laws that prohibit or regulate the ownership of primates. States such as California, New York, and New Jersey have harsh bans, although others require permits and high care standards.
- Internationally, countries have varied restrictions for exotic pets. For example, in the United Kingdom, the Dangerous Wild Animals Act requires a license to acquire primates. Primate possession as pets is largely prohibited in Australia.
- Before selecting a macaque as a pet, you must first research and understand the applicable legislation in your area.

2. What are the main health hazards associated with owning a macaque monkey?
Owning a macaque monkey carries substantial health concerns due to zoonotic illnesses, which can be passed from animals to humans.

- Herpes B Virus: One of the most dangerous threats is the Herpes B virus, which can kill humans if spread. The virus is primarily transmitted via bites, scratches, or contact with the monkey's saliva.
- Tuberculosis: Macaques can transmit tuberculosis, which is highly contagious and spreads by respiratory droplets.
- Other zoonotic diseases include the simian immunodeficiency virus (SIV), hepatitis, and numerous parasitic infections.
- Regular veterinary care and strict cleanliness procedures are critical for reducing these dangers.

3. What do macaques eat?

Macaque monkeys have complicated nutritional requirements that must be supplied precisely to preserve their health and well-being.

- Fruits and vegetables: Fresh fruits and vegetables, such as leafy greens, berries, apples, and carrots, should make up a large part of their diet.
- In the wild, macaques eat insects and tiny animals. This can be supplemented in captivity with boiled eggs, prepared lean meats, and specially made primate pellets.
- Foraging Opportunities: Giving them opportunities to forage is critical for their mental stimulation and well-being. This may include hiding food in toys or scattering it over their area.
- Obesity, malnutrition, and metabolic bone disease are all preventable health concerns that require a balanced diet.

4. How long do macaques live?

Macaque monkeys can live for 25 to 30 years in captivity with good care. This extended longevity necessitates a tremendous commitment from the owner in terms of time, finances, and effort.

5. Do macaques have to live with other monkeys?

Yes, macaques are extremely social animals who thrive on association with their own kind. Keeping a solitary macaque without companionship might cause serious behavioral and psychological problems.

- Social groupings: In the wild, macaques form complex social groupings. Mimicking this social milieu in captivity is critical to their mental wellness.
- Behavioral Enrichment: Without social connection, macaques may exhibit deviant behaviors such as self-mutilation, aggressiveness, and depression.

6. Can macaques be trained?

Yes, macaque monkeys can be trained with positive reinforcement approaches. Training can help manage their behavior and make it easier to care for them.

- Training can involve simple orders like coming when called, displaying a body part for examination, and entering a container.
- Behavioral Management: Training can also help you manage harmful behaviors, relieve stress, and provide mental stimulation.

- Consistency and Patience: To train macaques, you must be consistent, patient, and understand their behavior.

7. What type of enclosure do macaque monkeys require?
Macaques require huge, complicated cages with adequate space for climbing, foraging, and playing.

- Size and space: The enclosure should be large, with a minimum size of 10 feet by 10 feet by 10 feet for a single macaque. Bigger places are usually preferable.
- Environmental enrichment should include climbing equipment, ropes, swings, and toys to keep the macaque cognitively and physically active.
- Outdoor Access: Regular access to an outdoor location provides exposure to natural sunshine and a more diverse environment.

8. What are the costs of keeping a macaque monkey?
The expenditures of owning a macaque monkey can be high, including initial setup, continuing care, and unexpected expenses.

- Initial Setup: The cost of obtaining or creating a suitable enclosure might range between hundreds and thousands of dollars.
- Ongoing Costs: Regular expenses include food, veterinarian care, enrichment items, and enclosure upkeep.
- Emergency Costs: Unexpected veterinary crises can be expensive, requiring specialist care.

9. How do macaque monkeys display aggression?

Macaques can express aggressiveness in a variety of ways, including biting, scratching, and vocalizing.

- Physical symptoms of aggression include bared teeth, lunging, and swiping.
- Loud screeches and violent vocalizations are frequently used in conjunction with physical aggressiveness.
- Body Language: Understanding macaque body language is critical for detecting symptoms of oncoming violence and taking protective actions.

10. What are the ethical considerations in having macaque monkeys as pets?

There are serious ethical considerations in keeping macaque monkeys as pets.

- Welfare Concerns: Because macaques have complex social, physical, and psychological demands, it is difficult to ensure their welfare in the home.
- Conservation Impact: The pet trade can harm wild macaque populations and conservation efforts.
- Long-term Commitment: To secure the macaque's well-being, owners must be willing to make a long-term commitment, which can last several decades.

11. Can macaques be potty trained?

Potty training a macaque monkey is difficult and not always successful.

- Partial Success: Some owners may have some success with litter training, although mishaps are prevalent.
- Behavioral Challenges: Macaques' inherent behavior does not lend itself well to consistent potty training.
- Hygiene Management: In order to effectively manage hygiene, owners must be prepared to perform regular cleaning and maintenance.

12. What kind of veterinary treatment do macaques require?
Macaques require specialist veterinary treatment from experts in primate health.

- Regular health check-ups are required to monitor for symptoms of sickness and ensure that vaccines are up to date.
- Emergency Care: Having access to emergency veterinary care is critical for addressing health issues quickly.
- Preventive Measures: Preventive care, such as parasite control and dental care, is essential for overall health.

13. How do macaques communicate?
Macaque monkeys communicate by vocalizations, facial expressions, and body language.

- Macaques employ a variety of vocalizations to express emotions, warn others about danger, and communicate their social standing.
- Facial Expressions: Lip-smacking and teeth-baring transmit a variety of meanings, ranging from friendliness to violence.

- Body Language: Understanding body language, such as posture and gestures, is essential for deciphering macaque communication.

14. What can I do if my macaque monkey turns aggressive?
If a macaque monkey gets violent, it is critical to take prompt action to guarantee its safety and treat the root causes.

- Safety first: Avoid direct conflict to ensure your own and others' safety. Use barriers or fences to keep oneself safe from the angry macaque.
- Consult a Professional: To understand and resolve the source of the aggression, seek guidance from a veterinarian or an animal behaviorist who has knowledge in primate behavior.
- Environmental Enrichment: Improving the surroundings and offering additional mental and physical stimulation can assist minimize aggressive behavior.

15. Can macaques get along with other pets?
Macaque monkeys do not get along well with other pets and can be dangerous to them.

- Aggression Risks: Macaques can be territorial and aggressive toward other animals, endangering both the macaque and other pets.
- Different Needs: Macaques' care needs and social behaviors are notably different from those of household pets, making cohabitation difficult.

- Safety Measures: If there are other pets present, rigorous supervision and separation are required to protect their safety.

16. How can I improve the life of my macaque monkey?

Enriching the life of a macaque monkey entails providing a variety of activities and stimuli that match their natural habitat.

- Foraging Activities: Providing foraging chances by concealing food in toys or puzzle feeders promotes natural behaviors.
- Climbing Structures: Offering climbing structures, ropes, and swings encourages both physical activity and mental stimulation.
- Social engagement: Regular social engagement with other macaques, or, in their absence, with human caretakers, is critical for their psychological well.
- Toys and Puzzles: Providing a range of toys and puzzles helps macaques stay active and avoid boredom.

17. What are the indicators of stress in macaque monkeys?

Recognizing indicators of stress in a macaque monkey is critical for meeting their requirements quickly.

- Behavioral Changes: Stress symptoms include increased aggression, self-mutilation, pacing, and withdrawal.
- Physical symptoms of stress may include hair loss, weight loss, and appetite changes.
- Vocalizations: Increased vocalizations or odd sounds may indicate tension or discomfort.

18. How much time should I spend with my macaque monkey?
Macaque monkeys demand a large amount of social interaction and attention from their caregivers.

- Daily interaction and engagement are critical for meeting their social and psychological requirements.
- Quality Time: Spending quality time with your macaque, such as playing, grooming, and training, strengthens the bond and relieves stress.
- Consistency: Macaques benefit from consistent routines and interactions, which help them feel secure and stable.

19. Can I travel with my macaque monkey?
Traveling with a macaque monkey is complicated and generally necessitates specific precautions and preparations.

- Legal Restrictions: Travel regulations vary by country, and transporting exotic animals is often restricted or prohibited.
- Stress and Comfort: Macaques can experience stress when traveling. Ensuring their comfort and safety while traveling is critical, which includes providing a secure carrier and familiar items.
- Health Requirements: Health certificates and immunizations may be necessary for travel, especially to international destinations.

20. What should I do if I am no longer able to care for my macaque monkey?

If you can no longer care for your macaque monkey, you must find a responsible and ethical alternative.

- Contact reputable sanctuaries or rescue organizations that specialize in monkey care to see if they can house your macaque.
- Rehoming: If rehoming is required, ensure that the new caregiver has the requisite knowledge, resources, and legal rights to care for a macaque.
- Avoid Pet Trade: Avoid selling or gifting your macaque to people who may not be able to offer sufficient care.

Keeping a macaque monkey as a pet entails a number of unique obstacles and obligations. Understanding the legal, physiological, behavioral, and ethical issues is critical to maintaining the macaque's well-being and the owner's safety. These frequently asked questions (FAQs) provide a detailed overview of typical issues and considerations, allowing future owners to make informed decisions about the challenges of caring for macaque monkeys.

Conclusion

Is a Macaque Monkey Right For You?

Determining if a macaque monkey is the ideal pet for you requires careful consideration of a variety of variables, including legal and ethical concerns, the intricacies of their care, and the impact on your lifestyle. While these primates can be intriguing and clever companions, they also pose distinct obstacles that prospective owners must thoroughly understand and ready to face.

Legal and ethical considerations

The decision to keep a macaque monkey as a pet is largely impacted by legal and ethical concerns.

- Legal considerations

Jurisdictional Laws: The legality of having a macaque monkey varies greatly according to your location. Many regions have tight regulations or outright bans on ownership due to concerns about public safety, animal welfare, and conservation.

Permits & Licenses: Even in areas where ownership is authorized, obtaining the requisite permits and licenses can be a complicated and time-consuming procedure. It frequently entails displaying an expertise of primate care, providing adequate housing, and abiding to stringent welfare requirements.

International Trade: The import and export of macaque monkeys is governed by international legislation such as the Convention on International Trade in Endangered Species of Wild Fauna and Flora (CITES). Violations of these regulations may result in severe legal penalties.

- Ethical considerations

Animal Welfare: Macaque monkeys have complex social, physical, and psychological demands that can be difficult to address in a home context. Ensuring their well-being demands tremendous resources, knowledge, and attention.

Conservation Impact: The pet trade may contribute to the exploitation and depletion of wild macaque populations. Responsible ownership entails thinking about the broader influence on conservation efforts and supporting activities that promote the protection of these species in their natural habitats.

Ethical Treatment: Caring for a macaque monkey involves enrichment, social interaction, veterinary treatment, and a suitable environment that emphasizes the animal's well-being over personal desires.

Care and Maintenance

Owning a macaque monkey involves a significant investment of time, effort, and money. Their care requires specific expertise and regular monitoring of numerous areas of their health and well-being.

- Housing & Environment

Enclosure requirements: Macaque monkeys require large enclosures that allow for natural behaviors such as climbing, foraging, and social engagement. The enclosure should be secure, well-ventilated, and filled with toys, perches, and hiding spots.

Outdoor Access: Having regular access to an outdoor location is good for their physical health and mental stimulation. Outdoor cages should be built to prevent escape and keep the macaque safe from any threats.

Environmental Enrichment: Enrichment activities such as puzzle feeders, climbing structures, and social engagement with compatible companions or caregivers are critical to their cognitive and emotional well-being.

- Diet & Nutrition

Macaque monkeys require a well-balanced diet rich in fruits, vegetables, protein sources, and specially developed primate pellets. Meeting their nutritional requirements helps them avoid health issues like obesity, malnutrition, and metabolic diseases.

Foraging Opportunities: Stimulating natural foraging habits with food puzzles and hidden treats promotes mental engagement and minimizes boredom.

Hydration: Having access to clean, fresh water at all times is critical for their general health and hydration requirements.

- Veterinary Care

Routine Health Checks: Regular veterinary checks are required to monitor their health, detect any problems early, and deliver immunizations.

Emergency Care: Having access to emergency veterinary services that specialize in primate care is critical for dealing with accidents, diseases, and unforeseen health crises.

Parasite Control: Preventative practices, such as parasite control and dental care, contribute to their long-term health and well-being.

Behavioral and Social Needs
Macaque monkeys are extremely gregarious and clever creatures who thrive on interaction and mental stimulation. Meeting their behavioral demands is critical to their overall happiness and well-being.

- Social Interaction

Group Dynamics: In the wild, macaques spontaneously form social groups. Providing opportunities for socialization with compatible partners or caretakers helps them achieve their social interaction needs while also reducing stress.

Behavioral Enrichment: Enrichment activities, such as training sessions, puzzle toys, and fresh experiences, boost cognitive development while also preventing boredom.

Encouraging play and physical activity through toys, climbing structures, and outdoor exploration benefits both their physical and mental health.

- Training and Enrichment

Positive Reinforcement: Using positive reinforcement tactics, such as rewards for desired behaviors, can help create trust, regulate aggression, and promote cooperative relationships.

Cognitive Challenges: Engaging macaques in cognitively demanding activities such as problem solving and sensory enrichment improves their cognitive ability and prevents behavioral disorders.

Consistency and Patience: Successful training and enrichment necessitate patience, consistency, and a thorough understanding of macaque behavior and communication.

Challenges and Considerations

Owning a macaque monkey poses various problems that prospective owners should carefully examine before making a decision.

- Health Risks

Zoonotic Diseases: Macaque monkeys can spread diseases like Herpes B virus, TB, and parasites to people. Strict hygiene and regular health monitoring are required to reduce these dangers.

Behavioral Issues: Aggression, territorial behavior, and stress-related behaviors might endanger both the macaque and its caregivers.

Legal Compliance: Complying with local laws, obtaining essential permissions, and satisfying regulatory standards are critical for avoiding legal problems and ensuring responsible ownership.

- Long-term commitment

Lifespan: Macaque monkeys have a lengthy lifespan, frequently lasting 25 to 30 years or more in captivity. Potential owners must be prepared to make a long-term commitment to providing for their requirements throughout their life.

Financial Responsibility: Owning a macaque monkey incurs significant costs, including veterinary care, specialist diet, enclosure maintenance, and enrichment, which must be carefully budgeted for.

Ethical Responsibility: Ensuring the ethical treatment and care of a macaque monkey necessitates constant education, animal rights advocacy, and funding for conservation efforts.

Making Informed Decisions

Determining if a macaque monkey is suitable as a pet takes extensive research, honest self-evaluation, and consideration of ethical considerations. While these primates can create special ties with their caretakers and exhibit fascinating behaviors, they are not domesticated animals and have distinct requirements that may be difficult to provide in a household setting.

Before purchasing a macaque monkey, potential owners should:

- Extensive research: Learn about the species, their natural activities, and the specialized care requirements for macaque monkeys.
- Evaluate Personal Readiness: Assess your ability to meet their physical, social, and emotional requirements throughout their lives.
- Consider Alternative Options: Look into other methods to assist macaque conservation and wellbeing without participating in the exotic pet trade.

Finally, appropriate ownership of a macaque monkey entails prioritizing their well-being, promoting conservation efforts, and making informed decisions that are consistent with ethical ideals and legal requirements. By considering these qualities, potential owners can assess whether a macaque monkey is a good fit for their lifestyle and degree of commitment.

www.ingramcontent.com/pod-product-compliance
Lightning Source LLC
Chambersburg PA
CBHW071939210526
45479CB00002B/740